SWING BEAM

SWING BEAM

My Father's Story of Life on the Farm
and the Barns He Loved and Lost

LAURA LUSH

Life Rattle Press Toronto, Canada

Swing Beam:

My Father's Story of Life on the Farm
and the Barns He Loved and Lost

by Laura Lush

First Canadian Edition

Published by Life Rattle Press, Toronto, Canada

Copyright © 2013 by the author.
All rights reserved. The use of any part of this publication, reproduced, transmitted in any form or stored in a retrieval system, without the prior consent of the author is an infringement on the copyright law.

Life Rattle Press New Publishers Series

ISSN 1713-8981

ISBN 978-1-927023-62-4 (pbk.)

Copy edit by John Dunford
Cover design and typeset by Laurie Kallis

For my father.

CONTENTS

Section I

Beginning: Barn Talk 11

Tell Me All You Know About Barns 31

Walkers Line, Burlington, Ontario 39

St. George, Ontario 41

St. George Barns 49

The Pig in the Shoe Store 53

Running and Farming 59

Running and Farming: Pan American Games 63

My Wife 69

When Things Burn 73

The Last Farm 83

Piggery 87

The Goose Man 89

Christmas Delivery 93

Cleaning 101

Looking Back 103

Owen Sound II 105

Drunk Piggery 106

Return to the Garden 113

Planting: May Long Weekend, 2012 115

Section II

Changes 123

Taking the Barns Down 139

A Man and a Dog 147

Acknowledgements 161

About the Author 163

LIST OF IMAGES

Section I

The laneway to the farm, circa 1976 before trees were planted. 5

Peering inside my father's barn. 6

My father carrying his pails of water to feed the pigs. 8

Close-up of a beam inside my father's barn. 10

Another glimpse of beams in the barn. 18

The frame of the barn. 20

My father dwarfed by the huge cavern of the barn. 22

The inside of the barn roof: 1. 26

The inside of the barn roof: 2. 28

An aerial view of my father, his geese, and Robbie McGlashan, one of his many dogs. 36

My father running in a marathon race. 58

Postcard my father sent to his parents as his plane was arcing over St. George. 64

Marion, one year married. 68

Feature article in *The Hamilton Spectator*, December 19, 1963. 72

My father and his friend sitting on the front porch steps. 78

Flossie with her piglets. 83

Sows. 86

Gosling and geese. 88

Geese photographed from the barn roof. 89

Feature article in *The Sun Times*, December 17, 1983. 92

Two of the last three geese bathing in the pig trough. 95

Dad captures the last goose to go to the Bumstead Homestead. 95

Farm fresh eggs lined up on top of the dryer. 98

One of the chickens peeking out of the chicken pen door. 99

Wheeling a load of manure out of the pig barn. 100

The Farmer's Advocate and Home Magazine. January 15, 1893. 102

My son, Jack, feeding the grey mare. 108

My father fixing the roof of the barn. 110

Dad's Garden. 112

Jessie and Jodi in the truck. 116

The hay mow. 118

Section II

My father's sailboat in the bottom of one of the barns. 125

My father walking into the empty barn, except for two TVs. 130

My father gazing at his old treasures taken out of the barn. 132

The pile of scrap wood taken from inside of the barn. 132

Inside of the barn, empty. 133

Fallen debris, including a For Sale sign. Lush Realty is owned by my uncle, Peter. 133

Workers start dismantling the barn. 136

My father walks past the barn as it is dismantled. 138

The demolished barn, photographed from the concession line. 141

Victorian concrete ornaments that topped the gate posts at the laneway entrance. 142

The house and the fallen barns. 144

Dad as a teen with one of his first dogs, Bud. 147

Dad as a toddler. 153

The barn, stripped of boards, still standing erect. 154

The silo flush to the side of the barn. 156

Cat in the empty barn. 158

SECTION I

We tell stories—the stories tell us.

How to Find Me

When you drive south on 21,
look for the barns. The barns,
they aren't shy. Big and grey, rising
out of the landscape like a hillock of timber,
like a clutch of trees.
(You don't have to justify a tree.)
(You don't have to justify a barn.)
There are cattle in those barns.
Pigs, rats, mice, cats. Even a couple of boats
I keep in the lower part of one barn.
A tomb of sorts. Treasures, memories, and
oh stories—I could go on.

Slow down as soon as you pass Concessions 10 and 11.
Don Scott's house is on the left, banking the corner field.
Don and Gwen, they have been our neighbours
through good and bad, through summers
and winters, through all the losses.
We were there when Don lost Gwen;
Don was there when we lost our son.
You'll see Don on his rider lawnmower—brown Tilley hat,
glasses—turning out of his lane, then down the side of the road
to mow our grass that fringes the long driveway.
That's the kind of guy Don is.
That's the kind of neighbour he is.

Across the road—the Walker farm. They keep trotters
and the Jack Russells that tore through
my turkeys one spring. I said if it happens again, I shoot.
Then I got my own dogs—Jessie and Jodi—they make sure
the Jack Russells don't tear through my turkeys anymore,
make sure I don't have to shoot.

Put on your left indicator, ready yourself for the long
drive up the lane. Keep driving past the house—wave
to Marion leaning over the kitchen sink or reaching
for a cup—keep driving past the chicken pen,
past the purple phlox, the Hawthorne bush. Crack your window
if you want—the whiff of growing things.
Don't mind the dogs. They just like to tell you
they are there. Get out and open the gate to the barn yard.
Step carefully. The geese—their tongues will fly
at you like Dragon Tongues. Park in front of the first barn—
up the rock hill, beside the hay wagon.

Stop there and get out of your car.
Lift the door hasp—with one hand—it will
slide easy enough. Walk in. Your first deep breath
of barn. Walk into the wild pigeon flap.
I'll be there, grain bag slung over my shoulder or
pitching hay through the chute.
Take your time.
I'll be there 365 days a year
for the next 35 years.
And then one day,
I won't.

The laneway to the farm, circa 1976 before trees were planted.

Peering inside my father's barn.

I Remember

I remember what I need to remember:

One wife, three children—some things

staying too long, others leaving too soon.

I remember too many dogs to name:

Bud, Chubby, Duke and Charlie, Lindy,

Lady, Robbie McGlashan, Rhondo,

my last dogs—Jessie and Jodi.

Mostly, I remember the land.

The land frozen, the land green,

the land too rocky, too wet, too dry.

Always working it—tilling it—discing it.

I remember trying to start tractors, chain saws,

water pumps. I remember trying too many times

to do too many things. I remember pulling combines

behind me, my neck crooked to the right, then to the left.

I remember driving those tractors back into the barns, how

quiet and dark in there. I remember climbing ladders to fix

the barns after storms, hail, tornadoes plundering.

I remember every morning, every afternoon, every evening.

I remember feeding chickens, turkeys, cows, pigs. Sometimes

shooting groundhogs, coons and foxes. Almost a neighbour's dog.

I remember going to my barns just to fix things—look up.

I remember I have never been a religious man—

yet somehow in those barns—a pulling up into vaulted beams.

I remember almost believing in something

I still don't know how to say.

My father carrying his pails of water to feed the pigs.

Morning

Wake up, slip on my coveralls,

my barn boots—a hat, a pair of gloves.

Then walk out to those barns.

Hundred-year-old-boards sawing in sleep.

    Barn anchored to earth,

        the earth anchored to barn.

I can't tell you how they do it—

shuck winter, bust winds, ping hail,

    tighten their beams.

Tell me what has more stories

    than these barns living?

Toward the Barn He Walks Every Morning

Two white buckets, one in each hand.

Water-slosh, boot-squeak.

The geese roiling. Jessie and Jodi, noses to the ground.

Because life once in those barns.

Cows lowing. Shimmer of calf.

He holds his hands to his eyes—sees 20 years ahead—

barn boards tossed. Men hard with hammers.

A game of pick-up sticks.

    Yet still walks toward

    not away

    from those barns.

Close-up of a beam inside my father's barn.

How They Built Them

We all know the story—
mallets and levers, mallets and pike-poles.
Men working, men working together.
The women feeding the men working together
and the children watching the women feeding
the men working together, and the whole damned
village pulled/keeled/fixed to one purpose. *Getting
the barn up.* The barn raised tall and timber-high.
Giant pined construct, thewly beauty of bevelled
mitres fit together so the hay, the straw,
the sheaves—so the cats, the cows, the
horses all inside, all thrusted warm and tight
inside.

Spend time on the land, spend time with the living, spend time with the living, not the dying.

Beginning: Barn Talk

THE FLOOR OF THIS NEW REC ROOM IS ORIGINAL SLATE taken from the Wiarton escarpment. A fossilized skeleton of an animal's back curves out of the slate in a *J* pattern. Around us sit unopened boxes of LPs and books and a maze of paraphernalia. Behind him, his running trophies line five rows of shelving, the gold bodies of men with their hands thrust up in the air, faces pained. Trophies missing hands or feet, or both—torsos of chipped bronze and silver lying on their sides. An old wood stove rescued from his recently-sold farm is mounted into the west wall of the rec room.

It's an unusually warm day, but my father wants the wood stove started. He balls up some old newspapers and reaches for a couple of logs, then leans forward in his black La-Z-Boy chair. He scrapes a match along the box and lights the fire.

He lights the fire because today he is going to tell me a story.

The first barn I remember is a very small one. It was the back of a cottage on the Beach Strip in Burlington, a combination of a garage, woodshed and barn. It originally had a couple of horse stalls. There were horses, wood for the winter, and if we were lucky enough—a car.

Pop's car was kept there. An old Packard. The interior floor had all rusted out and Mom wouldn't let us put our feet down. She was afraid we were going to fall through the holes in the floor. I remember looking down and watching the road zip by, my feet curled up on the seat. That's the first barn I remember seeing. I was four years old. I had some baby ducks—Pop gave them to me—and I had some dogs. Two.

The next barn I remember was on Guelph Line; Pop rented this farm. It was three and a half acres and it had a really nice barn. You drove in and it had a garage on the first floor, chicken pens upstairs—Pop raised chickens for a living—and on one side, a mow to put your straw in. It was all wood. It looked quite sharp. It had been well-looked after. Probably built in the twenties or thirties.

The next barn I remember is on Torrance Street. I was nine or ten years old. Just prior to the war. There were two big buildings. One big, long shed they kept their baskets in—cause it was a market garden—and a big barn. A bank barn. This particular barn was in terrible shape. Wasn't worth fixing. It had been let go too long, so Pop tore it down.

The next barn was in 1940. Pop had bought this brick home on Brant Street next to the public school and it had been a farm originally, as almost all houses on north Brant Street were. Apple farms, vegetable farms. This was a really nice barn. It had a cement foundation. The first floor was all stone and there was a big hay mow above. It had a steel roof—a tin roof—and one section had been set up as a garage and you kept your car in there. There was also a spot where they had kept their horse and buggy. I kept chickens in the barn, and some ducks, some pheasants, and up to five dogs at a time.

There was a rule in our house that I could have as many dogs as I wanted—as long as I looked after them. The day I didn't look after them—they were gone. I knew a dog catcher named Clanky. He would always call me and he would say, "Have I got a dog for you!" I would go down and I couldn't resist. So to pay for the dog food, I pedalled eggs all over Burlington in two six-quart baskets, and I'd sell them for so much a dozen, and people liked them 'cause they were all fresh, a few days old.

The next barn—I'd be a teenager then. I finished high school and we bought a sixty-acre farm on Walker's Line and it had a massive house, a stone home. It was the original house on Walker's Line and it had a massive bank barn, which was in bad shape. I reroofed it, patched it up. One wall at the far side had caved in, so I took out the loose stones, and we cemented them up and filled them up again. It had four horse stalls, a couple of big pens for pigs, and a small area for cattle. That was a typical bank barn. Built after 1880. Bank barns were built so you could just pull up your wagon and drive in. You just drove up into it. To get to the upstairs, you didn't climb a ladder, you went around the side and walked up a ramp—a bank ramp. A lot of these barns were built into a hill. If not, they brought earth and built it up on one side. So you'd take your hay wagon with a team of horses and pull right in.

After that farm, we had the farm in St. George, a mixed farm. A family farm. It had a little bit of everything.

Then the farm in Owen Sound. My last farm. I was there for thirty-five years straight. There were two barns. One strictly built for horses. It had a high ceiling. The first barn. The best barn. And the next barn was for cattle. And hogs. I wanted to go into the sheep business as well.

Here's a picture of the second floor of the barn. It's got beams. Beach logs cut from the bush with a saw. You'd take an axe and square it off. It wasn't just boards, you understand. It was a piece of a tree brought in by a horse. One piece of tree at a time. You'd cut it up to size and fit it onto the cement wall that was built to the stone wall that covered the first floor. You'd fit in the notches, and then you'd put the rest of the beams up. There were almost no nails used. No nails. They all fitted into notches. There would be one coming up this way—and one coming that way. There would be a notch, and you'd cut it into here and there. And they would fit in *exactly*.

This is where you come in with your machinery and the hay mow is up above. Bank barns were built for a style of farming that was done at that time and it covered a long space. They were built from the 1860s right up until twenty, thirty years ago. They didn't build barns like this after that. The only ones who do are the Mennonites because they're still labour intensive. They don't have tractors. They have horses. They plow with horses. They bring in their hay loose. They farm like it was the turn of the century. And they have big families. They help each other. They're in communities. When George is haying, they all help and bring it in in one day. And the next day, George helps the other neighbours. They build barns exactly that way. They do their crops that way.

But we never did. Protestant people. The settlers did long ago, but we got more independent. When we built barns—sure, there would be two or three neighbours who came along, but you didn't necessarily go next door and help your neighbour bring the harvest in. But the Mennonites do that. You have to do that when you have no machinery. Machinery short-circuited the farming operation by allowing one man to do a tremendous amount of work if he had a good tractor and good equipment. People don't want the physical work they had to do before. That's *hard* work.

The Barn Speaks: I

Tenoned and pegged,

Morticed and boxed.

    I rust no nails.

    I am let into each post.

Driven hard.

Driven hard so wind can't break me.

Driven hard so nobody can

    take me down.

Another glimpse of beams in the barn.

The Barn: Etymology

The ere.

The bere.

The byre.

The bryne.

The *barn*.

The frame of the barn.

Early Barn

Wattle—a criss-cross of twigs.

Daub—the thickly mud and straw

plastering the wattle down.

A roof, thatched straw. No wolf

dare blow down. A narrow hole

at the top so smoke could escape.

(They lived in their barns.)

On those walls, a few hooks to hang

their things—rough coats, hats, mitts.

A wooden table, a couple of chairs.

The stench! A bed of straw. And more straw—

ricks for hay and straw—the sheen!

Cattle stowed below the parapet.

Lowing them to sleep each night

in these U-shaped crucks.

My father dwarfed by the huge cavern of the barn.

How to Build a Barn

You take two trees. Trim their branches,
roughly.
    Bent together, they will fall like men
falling head to head, exhausted from fight.
And they will form a beginning—an *alpha*, an A, a frame,
    a *cruck*. And another pair of trees again.
The same. Braced by a single pole.
    Ah, men,
drop your axes. This is all you'll ever need.

Build

Build with your hands.

Build nail-free.

Build with what's around you.

These twigs, these roots, these branches.

Build with beach and maple and oak.

Build until these trees are rooted again,

homed in this basilica.

Let the trees swing their beams

through the great wide space.

    Collar beam.

    Tie beam.

And just let it all alone.

Framework

These spaces between the beams
and posts. The hay, the straw—
breathes. And the barn's lungs open.

The inside of the barn roof: 1.

Swing Beam

Heavy-timbered,

35-foot beach swing beam

spanning the barn's interior.

No centre post on the barn's threshing floor.

Tether ox to a pivot. That ox will circle,

thresh grain. At the centre, the bay.

And during the winter, the weight of the winter

will crush down on that swing beam.

Bearing, singular. A sturdy truss

that supports the loft and

holding that hay-girth in.

The inside of the barn roof: 2

Threshing Floor

Honey coloured.
The wagon is full.
Into the barn, take wheat.
Flail it hard. Until the wheat softens,
until the floor shines.

Memory at Seventeen

We would walk up that hill,
the barn door padlocked. Just dark,
behind those doors?
Slide the lock, pigeons' coo.
The all-at-once caught
of something in the throat—and
then the immensity,
the hand-rough of those beams,
hay tilting, leaning, the barn laughing
out its sides.

Tell Me All You Know About Barns

A THATCHED BARN. NOT TOO UNLIKE WHAT YOU would see in England, and it was used out west. That's because they didn't have all the wood that we have here. It was a mudbarn as well, because they filled in the cracks with mud and it baked. I've never seen one alive today. Never saw one as a kid either because it was out west.

Then you had barns with a truss roof. This means that you have posts up there and strips across here and there is a peak roof, and what it does is give you the space to put your hay in and build right up to the top. You'd fill the barn right up to the roof. You'd pile the hay right from the floor. You'd drive the wagon in, and you'd fill it right to the top. And you'd fill the other side to the top. That's how I did my barns. I love the shafts of light the most. You could see the sky through those boards. Barns were made with wood cut locally and it hadn't dried out completely, so they'd nail it on—at that time the boards fit really tight together—but they'd dry out and after a while, you'd have half an inch gap in-between the boards.

Now, the typical Ontario barn. The bank barn. It's built into a hill so that your wagon can get up to the main floor, which is upstairs, very easily. You drive your team of horses in with your wagon and a fork would come down and lift your hay up. When I was doing it, I did that the first couple of years at St. George, but after that, I got a bailer. You'd just unload the bails and put them on a bale elevator ladder. It was a machine, with a chain that had hooks every six inches, powered by an electric motor. I would lay the bale on the elevator and it would go up to the ceiling.

You had to have a place to put your horse harnesses, and each horse had its own separate harness, and if I was hooking up Charlie or Vicky, then this would be Charlie's harness, and this would be Vicky's harness, and you'd keep all the bridles together.

You never want your barn to burn. You never want anything to burn. I had a burning barn at St. George. It was more of a drive shed—but it functioned like a barn—and I had turkey pullets under a heat lamp—an oil heater—and it caught fire and I lost them all.

The Dutch barns. I know these well. Wagon doors are placed at either side. But my barns were bank barns. They only had an entrance at one spot. They don't use nails. The wood is cut into a notch, with wood pegs shoved in to hold it tight. They used the wood of the area, maple and beach. A ramp goes up to the top of the barn. And there's usually an undershed where the horses could go when it rained.

You could put your tractor in there. That's where we put Trigger, the pony that died later—went to the barn to die.

Stone barns are unusual. They're a bit cold. A lot harder to build. I prefer a log barn—you can see the axe marks. They lay the logs on the ground and use an axe.

That's how the beams in my barn were. Cleaned up by an axe. Beautiful. The carpenter leaves his signature—his particular mark—right in the wood. They would cut these logs out on the ground, on the land, and they'd mark them. They'd say to themselves—this one is going to go here, and this one is going to go there, and this one is going to go across.

A typical Ontario barn. You can see all the beams. The hay on top. The mow separates the bay from the aisle. The mow is where they stack the bails. My barn had all these—the hooks and the hinges. Big, long hinges for support. They were all iron, from the turn of the century up to the twenties and thirties. Threshing and mowing. I never went to a barn raising, but I heard enough stories about them. The roof was a standard V-shaped roof with a thirty-degree angle to give extra width. It was just straight up and down.

We had a cement silo. They started out with stone silos, where they built a circular stone wall. They were very labour-intensive; they weren't normally very high because you had to put scaffolding up. They went from there to poured concrete where you'd bring in the forms and just pour the cement in. It was easy. They ran up a ladder along the side of it, poured the cement in, and they'd keep packing it in. The cement had to harden. My silo was a poured concrete silo. The smell of corn ensilage—you could get dead drunk in there.

In the barn, the rafters all hook in and strengthen each other. The cross pieces would hold the roof up, yet they would tie into the crossbeam, and again, no nails were used. They were notched in. They drilled a hole and they put a peg in; all my barns were built that way. The one at Walkers Line was like that, the one at St. George was like that. Ontario barns, built in the 1900s. My barns at St. George were built at the turn of the century, and they burned down in 1935, so they rebuilt them. What most people did was take the wood right out of the bush and they'd get a portable sawmill to come in on a wagon and they'd cut the beams to the measurement that they needed for the barns. The beams would sit for a while because the wood was green. And then they would dry out. It would save them from cracking and twisting when they put the barn up.

The Barn Speaks: Old English

Barne. A knafe barne.

That knafe barne had bene

Til knaf barne madyn.

Barne foure score,

Then teran down.

An aerial view of my father, his geese, and Robbie McGlashan, one of his many dogs.

From the Roof of the Barn

From the roof of the barn you
could see everything—
    The baby turkeys, bald and thin, the
gate I was trying to fix.
    A log waiting to be split.
Robbie leading the geese out or away from the barnyard.
    He could lead, that dog. Once my
son got right up there
    Just to take a picture of how I looked lugging
a pail of feed to the barn,
    geese plumply scattered.
And, Robbie, black and matted squatting
    the grass down.

You start with something, then it grows.

Walkers Line, Burlington, Ontario

I STARTED FARMING ON WALKERS LINE. I HAD A MARKET garden, sixty acres, no tractor at that time, so I went out to buy a team of horses. I knew this chap who would ride up and down Guelph Line with his team and he sold me the one horse but wouldn't sell me the other one because it was too nasty. If that was nasty, then God help us, because the one I bought would kick the hell out of you if it had the chance. His name was Tony. Tony would back off in the harness. Tony was quite a character.

I had a feed man who made rounds every week from Toronto and he dropped off feed—in those days the easiest way to do that. He said he knew of a horse who was really quiet, worked hard. I can't remember his name now, but he was a work horse. He really worked.

We moved to Walkers Line the first year after high school. Len Wilkensen was a neighbour who lived next door from us. He was always growing this and growing that and breeding this and breeding that and he talked me into breeding these mice. I was twenty. He had these pens, but he decided not to do it himself. He had stacks of tiny, little cages. This was not too long after the war, and there was a lot of research going on and these were special clinic mice, so this guy came in with a truck and dropped them off. It was a government-run operation and it was located in Buffalo. So I took these sixty mice and I bred them. It only takes three or four weeks for mice to be born, once you breed them, and I had mice coming out of my ears. And this truck drives in, an insulated truck from the States, and they picked up my mice and I got a nice cheque. I thought that was just great. I was through the next crop, because I bred them all at once, that way the truck wasn't coming in every week.

I got a registered letter saying that the government had withdrawn all finances for this research, so they wouldn't be needing any more mice. What was I going to do with all these friggin' mice? I took them to the market and sold them to some people who wanted white mice as pets.

What else could I do? Drown 'em? I was getting a buck a piece for these mice because they were a special breed of mice where the blood type was all the same, and they needed them for research, they needed the same bloodline. I tried to sell them anywhere at all where people wanted pets, but I ended up dropping most of them in the water pail.

St. George, Ontario

POP, MOM AND I MOVED TO ST. GEORGE—A VILLAGE OF eight hundred people six miles south of Galt on Highway 24—in 1952. I bought the team of horses that were there: Charlie and Vicky. Charlie was maybe two thousand pounds, real big. Dappled grey. Beautiful horse. And they could work. They worked good and steady.

Pete, my brother, would start out with the tractor and he'd plow and I'd come along with the harrow and cultivate it with the seed drill. A harrow had grate and teeth and you'd drag it along the field and break up all the lumps and make it nice and smooth and bring up all the stones to the surface. Peter would start out and Charlie would follow him. It worked really well. We did a lot in those days.

The barn we had in St. George was a nice barn, a bank barn. I painted it red by brush. I had to keep moving the ladder every five minutes. I kept at it. And when you're young, you can do things like that. I did it whenever I wasn't busy with something else. So my barn for half a year was half-painted, and finally I got it finished. I made it look really nice.

I decided I needed an income, so I went into the pig business. I was in my mid-twenties—twenty-five. I bought twenty-five sows. That seemed like a lot at the time; when I came up to Owen Sound, I had fifty sows. In those days, twenty-five sows were a lot of work too. I fed stock in the morning, fed the pigs and cattle, the chickens. I got up at six. I raised the pigs for market. I had a Danish barn. I fed them all in a trough with water, the old-fashioned Danish way, and I'd get good grades on the pigs.

I was also working in the fields. Full-time farming.

I had a team of horses and we'd seed everything down. I'd bring in the hay. Load it on the back of the wagon. I'd have to be on top of the wagon with the team of horses. I'd keep moving around and around until I got the load real high. I'd take it back to the barn, drive up the ramp, go into the barn. A big fork would come down and I'd stick it into the hay—it would lift the hay right up to the top.

In the fall, I was harvesting. Combining. We used to bag all our grain. We'd drop it down through the hole in the barn floor and grind it up to a mash.

You had the spring when you plowed, disced and seeded. You had the summer when you brought in all your hay. You had to fill the barn right up to the top. Late summer and fall, you harvested. Come winter, that's when you'd really work 'cause you were cleaning manure. You never ran out of work.

You worked seven days a week. It was a stock farm, livestock. And if you didn't stay on top...

I ran cattle every year as well. I'd sell the calves off, and then, if a cow acted up— aborted or I couldn't afford to keep it any longer, I'd sell it too. I sold a lot of wood, cut timber in the bush. And that was my income. I could throw out the manure on the raised area.

Cows are pretty smart. I would let them out in the morning and they'd head back along to the pasture field. I would drive the cattle out to the back and I'd leave them for the day and I'd go back and get them in the evening. The cows would recognize me. As soon as they saw me, they'd start coming back in. They knew they were going to get their grain.

I had a cow called Sue. You could ride her back. I'd just climb on her back and she'd walk in—very slow, very easy—and they'd all trail in, single file. Some of the cattle were still coming out of the bush when I'd herd the first bunch into the pen and feed them grain, a little bit of hay, a little bit of whatever you had.

Sue was so easygoing. But she rolled on a calf one day, just a couple of days old, maybe it was her fourth calf, and I got mad and I shipped her out. And I was sorry I did that. You know how you feel bad about doing things? I felt bad. I could've waited another nine months and she'd have another calf. But anyway, that's the way it goes.

It's a four-season job. Spring seeding, summer hay, fall harvesting, and the winter cleaning.

I was very careful and I would always treat my cattle so they didn't get pinkeye. In the summertime, you had to watch out for pinkeye. It was a type of bug that would infect their eye and they'd go blind. You had to put ointment on their eye. Try putting ointment on a cow's blind eye. They knew there was someone there, but they'd panic, and run into a tree. I'd have to go out and try to lasso them, put this paste on my finger. You could cure it, but if you didn't, you'd lose the cow. It was viral. It was terrible. And if one cow got infected in the herd, all the cows got infected. The flies brought it to the herd. Some years, they got infected; other years, they didn't. I was trying to treat this cow, it stepped on me, and then it crashed into a tree. I got up, and I had broken my ribs.

I had some great animals. I loved my animals. I had three or four good horses. Big draft horses. A tonne a piece. They were big and they could pull that seed drill or cultivator all day. I used the tractor for jobs the horses couldn't do. The tractor sometimes got stuck in the fields, so I used the horses. I could walk a team of horses through wet spots, and they dried it up in no time.

I never had a job I liked better.

———

We had some great parties in St. George, both in winter and summer. The kids were my age—twenty, twenty-five. They were all city kids from Toronto, Hamilton and Burlington, and they'd all come up.

In the summer time, we'd have a wiener or corn roast back at the lake—an eighteen-acre lake—and I'd hook the team up and we'd put a lot of hay on the wagon, bales of hay, they could sit on. They'd load up until we had a wagonful of kids, and then we'd head out.

I had two big Persian horses chomping at the bit wanting to go. We'd walk nice and sedately until we hit the bush road, and then it would go down a slope, and I would let the horses start to jog. The horses would start to get away and the girls would be screaming their heads off and the hay bales would be falling off and we'd be getting close to the lake. There was a road in the lake—you couldn't see it—you'd have to take your tractor to get to the other side. I'd get to the top of the hill and I'd let go of the reins and the horses would take off. Everyone was screaming and yelling, yelling and screaming, and I turned around and I'd lost all my passengers—because they thought they were going to get dumped in the lake. I'd be laughing my head off. The girls were all covered in mud, and wet, so everyone went for a swim that night.

In the winter time, I had a lot of parties. We really partied it up, as young people do. I had the place for it, a big place. Mom and Pop didn't mind. I asked a bunch of friends from Toronto to come out to the sugar bush. I was out making maple syrup. I had a great big steel pot, a big cast iron pot, and I had all the wood stacked up and the cans were hanging up along the trees. I'd empty the pails into the pot. I had it really going that day. I was out all day. They came out and I brought them out on the wagon and we shovelled and trampled down the snow and what we did was, we'd dip in with a cup and pour it on the snow and it would freeze and it would be sugar maple, like toffee. They thought it was great.

That farm is still there. I know where it is. I'd like to stop by and walk back. We left on good terms. The same people are there, but the wife died. It's now a Boy Scout camp.

The owner was a contractor and he tied in with a old guy in Toronto who had a contract with the first subway in Toronto. And darned if the old guy didn't die. And so he inherited this great big company. He went to the Olympics a couple of times, trapshooting. But when he bought the farm, he didn't want to shoot anymore. He just wanted to raise birds. He didn't hunt. He had a bunch of birds, pheasants, just like me. He was a nice enough guy. I didn't get a long with her too well.

I moved Pop and Mother. I had a tonne truck and I moved them and our stuff to the farm in Carlisle; it was a lot of work. I didn't have all the stuff off the verandah and she said this all has to go by four p.m. So I went over to the McCombe's and said I need some help. I had the house pretty clear, but I still had a lot of stuff on the verandah—beds, couches.

When we moved up there, I had a two-tonne truck to haul the tomatoes. I grew four acres the first time and we had a blight—the first time in twenty-five years—and most of the tomatoes were ruined. So the second year, I grew six acres.

Mr. Funston came around to see me and said, "Are you gonna stay with it, Barry?" and I said, "Yeah," and he marked me down for four acres. I said, "No, I want six acres," and he said, "Sure," and it was the driest goddamn summer and I barely made my expenses. Didn't make a profit.

So the next year, he came around and he said, "What are you gonna do, Barry?" And I said, "I'm gonna grow eight acres." And he said, "That's a lot, Barry." But I had to catch up.

Mom helped me a lot. I disced it up nice and smooth and I had a two-by-six and I put pegs in it, three pegs, three feet apart, and I dragged down one end of the field and back, and then I went the other way. It was like a checkerboard and you planted the tomatoes where the lines crossed. I had the earth worked up really good. I'd take a spade and work it into the ground. You put your foot on one side and pack it down. I had a crop of a lifetime. Was it ever good. All my friends from Hamilton came out and helped. All the guys who worked with me at the Brant Inn—Sandy, Doug and Don—and we kept up with it.

Anyway, the three places I farmed was Walkers Line, St. George and Owen Sound. So I had three farms over the period of, well, let's see…1950 until 2011. I was off the farm for ten years. We moved to Burlington in 1966. I was three years at Walkers Line, I was ten years at St. George, and thirty-five years here. But looking back over my life, I'm glad I did it. It wasn't the place where I made a hell of a lot of money, but I had a living.

St. George Barns

AT. ST. GEORGE, WE HAD 238 ACRES, A LAKE AND A FARM, an ideal stock farm. The thing I remember most about this farm was the Danish-style hog barn. It was a style of barn where you had big cement troughs and you mixed the mash with water the night before and you fed them the mash. The pens were set up so that you could isolate the pigs while you cleaned the pens and threw the manure into the barnyard. Each pen carried twenty pigs. This style made it very easy to clean a pen out because you didn't have to move the pigs out of the pen, you just shut the door and shoved everything down to one end, put the bedding down, and shoved the manure out through the exit door. It was a really slick way of doing things. I had a lot of people who had never seen a Danish-style barn before and they were really impressed.

I was working in the fields as well, full-time farming, a team of horses—just a great pair, Charlie and Vickie—and we'd seed everything down with grain and hay. I'd bring in the hay, load it on the back of the wagon loose. I didn't have a bailer in those days—just a hay loader—and you just drove over the cut hay and lifted it up and forked it around. When you got the load high enough, you'd go back to the barn and take it up the ramp where a big fork would come down from the top of the barn, and then you'd shove it right into the hay. A big steel fork with two pins and you'd shove it right down until you couldn't get it any further, and then you'd pull the latch which sent out teeth and then you'd lift it up. It came with the barn. You'd lift the hay right up to the top and then you'd drop it. Haying was the biggest part of my job.

Seeding went pretty quick. You plowed the field, you disced it, you harrowed it, and you smoothed it off. A harrow looks like a big gate lying on its side with pins on its side. You pull it and it breaks and smooths out the lumps of earth, and it will also lift the stones up so you can clear the stones up before you seed.

In the spring you plowed, disced and seeded, then in the summer you'd bring in all your hay. You had to fill the barn right up to the top. Late summer and fall, you harvested the grain. Come winter, that's when you really worked 'cause you cleaned the pens and fed the livestock, and it was a job that took you pretty much all day. You never ran out of work. You worked seven days a week. You had to feed seven days a week.

In St. George I had a big stable for cattle, with cow stalls where you tied them up. I took them out and made one great big pen because I wasn't going to have any milk cows. I had beef cattle. You could clean out the stalls very easily, if you had one big pen, with a front end loader. I brought in fresh straw to keep it nice and clean. If you had stalls themselves, you had to shovel it out with a fork.

I decided to change the horse stalls around too. There were five stalls, room to hang a harness up and one pen. I changed it into four standing stalls and two big loafing pens where you could put a mare or a colt. Frank Loan—my brother's father-in-law—disagreed with how to keep the horses. Frank insisted on having box stalls where the horses ran loose. But I wanted to keep them in stalls because they are so much easier to clean up. I'd just wheel in with a wheelbarrow in the morning and clean up all the manure, which was at one end of the stall, instead of having it all over the place. It made it easier for me, with work horses. It made it a lot easier to keep them nice and clean. So anyway, we got into a bit of an argument. I tore down what he had built and put it back the way I wanted it. I said, "Frank, this is still my farm. I appreciate all the help, and I look after your horses. So I think it's a reasonable tradeoff. Don't jump to conclusions and say you want it your way, because I pay the taxes."

So, we got that solved.

The Pig in the Shoe Store

I HAD SOME REALLY NICE SOWS AND I KEPT THEM longer than anyone else. I ran them out to pasture, so I would get eight or nine litters out of them, instead of just four. And some of them got pretty big—about four hundred pounds. I had big sows. They were quiet sows because I treated them right. I didn't kick them around or hit them, but this one sow went crazy in the heat when she farrowed. She smashed the door down in the barn, marched down to the pond, and had all her pigs there. When I heard the squealing, I went down there, and all the little pigs were lying belly up in the pond.

I was just furious. That was a big loss for me. I feed the sow all this time—four months—and she had eleven pigs and they drowned. She didn't know any better. She was lying in the mud. That would have been on a Tuesday.

I put the hose on her the next day and cleaned her up and I said, "I'm taking you to Kitchener, the stockyards. I'm not taking any more of this." So we loaded her up in an old beat-up pickup truck we had and took her to Kitchener. I had to go through Preston.

The sow didn't like being in the back of the truck. It was a Chev pickup truck with wooden racks. All of a sudden, I heard this crash. I stopped at a stoplight in Preston, looked, and said, "Holy Christ."

The sow had lifted up the back of the wooden frame that held her in and she's running down the main street of Preston. She was really running too. She was panicking.

I stopped the truck and I ran after her. She dodged between two parked cars and up onto the sidewalk. People were running to get out of the way. The more people yelled, the more frantic she got. So she ran into a shoe store. It was the summer time and the shoe store was really busy. And she ran in there and there was a whole pile of women in there—at least eight or nine—and two clerks, who were also women, and this great big thing like a goddamn elephant came tearing in and squealing.

Well, the women were screaming and running out of the store in their stocking feet and the sow ran into the back of the room. And if you were ever in a shoe store, they have these racks where the shoes are all stored, but the racks are very flimsy. She crashed into one and took all the boards down, pushed everything down, boxes flying everywhere, shoes pouring out. I was standing there with my mouth open. I was in a panic.

Just then, the sow decides she's had enough of this shoe store and she ran back out to the street and made a right hand turn down this alley. It was the only alleyway in Preston that was a dead-end, so it had her boxed in.

Two cops arrived right away. One cop reached for his gun and I said, "What are you gonna do? Shoot it? You can't do that." He looked as nervous as hell. He didn't want to have anything to do with this goddamn sow. And I said, "Just stand there and wave your hands when she starts coming toward you. I'll get my truck and back up in to the alleyway." And one cop said, "I'm not standing here. I'll help you with the truck or direct traffic." And the other cop had to stand there and wave his hands.

But the sow was so scared that she stayed in the back of the alleyway. So I backed the truck up as best I could. By now we had a big crowd watching. I backed it up and I dropped the tailgate down—it was terribly high off the ground—and the sow took one look at that truck and she ran to the truck, panic-stricken. It was the only thing she recognized. She jumped up with her front feet and couldn't get her back feet up. And there's two cops and myself with her back feet trying to lift this four-hundred-pound sow onto the truck. Got her in. I latched the gate down and tied some wire on it—I had a little bit of wire in the front of the truck—in case she tried to lift it again.

The one cop said, "Where are you taking her?" I said, "To the stockyards." Then the other cop said, "Well, we want her out of town. You'll have to come back because you'll have to answer for this." So I said, "I will. I promise."

They took my driver's licence number and everything—my address, my name—and I took off. When

I pulled off, there was a whole pile of people standing around the shoe store looking at all the carnage.

The sow never gave me another problem. We drove to the stockyards in Kitchener, to the loading ramp. She climbed off, and I signed for her, gave my name and address, so they could send me the cheque. I didn't hang around. I'm in a real mess. I drove back to Preston. I went to the police station and they said, "Well, you're going to have to see the owner; he's in Toronto, but he's coming back. So he'll charge you."

I went in and he wasn't there. I talked to the woman behind the counter and she was laughing. "I have worked in the shoe business for thirty years and I have never had anything like this happen before. I thought the damned sow was going to run between my legs. So I jumped up on the stool."

I said, "Well, tell the owner I'm so sorry and I'll come in in the morning and sit down and figure out what the damages are."

Well, the Kitchener-Waterloo paper and the TV station came out and took pictures of it. I was just relieved that I wasn't here, thank God.

I showed up the next morning, walked in the store, and the owner had come in to clean up and sort out all the shoes—'cause they were out of their boxes. He looked at me and said, "So you're the guy who wrecked my place."

I felt terrible. "I'm so sorry," I said. "It was one of those things that happened." He started laughing.

"It's okay for you to laugh," I told him, "but I'm the one who's got to pay for it."

"I tell you what I told the head office, Savage Shoes," he said. "They got so much damned publicity that we'll just forget about it."

My father running in a marathon race.

Running and Farming

I LOOKED AFTER THE ANIMALS; RAN MY SOWS OUTSIDE in the grass at St. George and I had a sow ready to farrow when I left for the British Empire Games. I said, "Peter, keep an eye on her. I don't want her farrowing on the grass." So Peter put her in the pen and she had eight or nine pigs.

I was at the Games for three weeks; it was difficult to leave the farm for that great of time. It was July, before the grain harvest. I remember writing to him:

> *"Dear Pete, by the time you get this letter, you will be well into the combining. Hope that everything is going all right and that you are not finding it too difficult to handle."*

Mostly, I was taking care of the farm at that time. When I was away, I worried about the farm—my family—and I thought seriously before going to the Games. Could I do it? Could I manage? Could they—Mom, Pop, and Pete—manage? Our whole life was wrapped up in that farm, and it just happened that I was going to the Games in July before the heavy combining. Wheat. We had more barley and oats than wheat, but by the time I got back, the combining was done.

It was 1954. I worked there till '62. I quit running in '58. I was married by then—1957—and that wasn't really fair to Marion. I'm gone half the night running down to Brantford and back. So, it was just a little too much. I was still running well and winning races, but I thought I'd better stop at the top.

Farming and running complimented each other. It allowed me to call my own shots as to when I'd train. I was up early in the morning with the livestock and ran in the night. I ran with army boots in the winter time to strengthen my legs. I'd run from St. George to Brantford. Managing chores and running. I trained three nights a week. I ran Highway 24. People knew me. They'd wave at me. I was young. I could take on a lot. Farming is a hell of a good profession if you're healthy.

During the British Empire Games, I met a girl out there who worked in an office—Eleanor, real friendly type—so I took her out a few times, but she was going on holidays, and she didn't stick around for the race. When the Games were over, we wrote back and forth, and she finally said, "I'm coming east for a holiday, and I'd like to see you." I said, "Okay." I wasn't interested in anything serious—a romance—we were just having fun.

> EMPIRE VILLAGE
> University of British Columbia
> VANCOUVER, B.C.
> CANADA
>
> Dear Pete
>
> I guess by the time you get this letter you will be well into the combining. I hope that everything is going all right and that you aren't finding it too hard to handle the chores.
>
> At the camp here I get up at 7 o'clock and eat breakfast, train till noon, and spend the afternoon walking around, resting, doing calisthenics and attending the coaches training school. The University of B.C. has a terrific model farm, 400 acres, run by their agriculture dept. I am spending quite a lot of spare time there and they certainly have a lot of new ideas in practice. I have heard a great deal from them about the Cariboo Region, an immense fertile valley in central B.C. which they claim is the greatest cattle district in Canada. They even offered to drive me up and give me a tour of the place but I had to turn it down as it would take about 3 days.
>
> I'll be coming home via United Air Lines out of Seattle probably on the 8th.
> Don't forget to send in the money on the Supertest Tires. It is due on the 26th July.
>
> See you soon
> Barry.

Letter my father sent to his parents from the 1954 British Empire Games in Vancouver.

So she came, and while she was there, the rodeo was on in Preston—a rodeo, now that was a big deal. The rodeo organizers came out because they wanted to borrow my horses—two great grey Belgian horses, Vicki and Charlie.

I said to her, "Okay. Why don't you come out and see the buck ride."

"You ride?"

"I don't think so," I said.

Eleanor arrived and she brought a girlfriend she knew out west. The girlfriend was going out with Al Utter—that's how I met this girl. So we drove out to the show to meet her, and I was anxious in case my horses acted up. Anyway, I got talked into riding the bucking broncos. Al said, "No friggin' way I'm getting on one of those horses."

I'd been living with horses for quite a while, so I wasn't too worried. I was wearing a cowboy shirt and the whole thing. I was full of bullshit, I think. I thought I was going to do okay.

It came my turn and I got on the chute. You stand on the rails and drop onto the seat. Of course, the horse doesn't usually do anything until you open the gate and then he comes out and he starts. Well, he threw me out of the stall. I went ass over tea kettle over the gate and landed in the rodeo yard.

They sure laughed like hell out of that. I was stunned. The horse didn't even make it out. I had just settled in the saddle and I was looking down to put my feet in the stirrups, and all of a sudden, it just lifted its rump and I went flying. Oh, it hurt. I hit my knee on the gate, rolled over on my side, and I rolled real quick because I thought that horse was going to come out and trample me, but it didn't. The crowd busted themselves laughing.

Running and Farming: Pan American Games

A YEAR LATER, I ENDED UP GOING TO MEXICO FOR THE Pan American Games. I still had a lot of responsibility on the farm, so it was tough to leave Mom and Pop. I had to run, I had to work some nights at the steel mill in Hamilton, and I had to farm. The farm came first, then the running, then the job at Stelco. But sometimes the running came first, especially for the big competitions. But I never stopped worrying about the farm.

Anyway, I was flying to Mexico, and it was around six-thirty in the morning. The flight left from Toronto, stopped in Chicago, then finally to Mexico. The plane takes a little time to get up to altitude. Three thousand feet, four thousand feet—you can feel the plane lifting. When it passed over St. George—I'd say at four thousand feet—I was looking out the window and I recognized Number 5 Highway. Then, I could see the village of St. George, and we crossed just a little left of the farm. This is the route you'd have to take to Chicago. The tractor was in the barn yard, but none of the animals were out, and I was always worried about the farm when I left. Pete, my brother, helped with the chores when I went to the British Empire Games the year before, but he wasn't available now. He was starting up his business. So Pop and I took over, but Pop didn't have to do any seeding or cultivating. He just had to make sure the hay was down from the mow for the cattle and feed the pigs—it worked out well that way. There he was, walking out to the barn. I could make out his figure from the plane window. I knew it was Pop. There he was walking to the barn. I know I said in the card that I didn't see him, but I did. I was joking with him. It would've been seven o'clock by the time we passed over St. George. That's the time Pop would be walking out to the barn to make sure the cattle had hay.

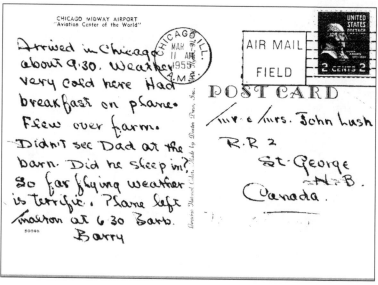

Postcard my father sent to his parents as his plane was arcing over St. George.

I didn't take a permanent job till '56. I didn't think my running days were over, but they were. I found that working all day and running half the night— sometimes ten, twelve, fifteen miles after supper—was a tough one to hang onto, but I enjoyed both the work and the running. I'd run from St. George to Brantford, about nineteen miles return, and I tried to do that three times a week. I'd race on Saturday. The running was my entertainment. I didn't mind working and I didn't mind training. If you really wanted to win a race, you had to train, so that's what I did. I had just ordinary women's light running shoes and the shoes they have today are fabulous—but in those days, they didn't have them. People knew me all the way up and down the 24 highway, and I'd wave to them and they'd blow their horns and wave back. My mother was a little uptight over all this because she thought I was working too hard and running too hard.

Pete left in '54, when I got back from the Games. He'd had enough. We started in '52. He said it was too much work for what you got out of it. You had to really like it. I liked everything about it. I liked growing things, raising livestock; I liked the independence of it. I didn't need a car. I drove an old pickup truck. And if the girl I dated didn't like the pickup truck, tough luck. But he left in '54. Pete hung in until I ran in the British Empire Games and then he went to work for my cousin's husband, Bill Hitchcock. Got his real estate licence and he never looked back after that.

I found that working as hard as I did still wasn't enough to pay the bills. So I took a job with Swift's, selling feed to feed mills because I knew the livestock business. I got a job there because I had been in the British Empire Games. I sent in my application and listed that I did a lot of running for my hobbies, that I was in the British Empire Games, that I finished fourth. That guy who interviewed me in Toronto for Swift's—a meat packing house based in Toronto—he was a sports person himself. He played basketball on a couple of teams. He said, "I'm gonna have this guy." The discipline that sports gave me trained me for a lot of things in life. So I got a job there and it worked out well. My boss was Art McFarlane; he hired me for the meat division, but I wanted to get into the feed division. So when I heard there was an opening, I went down to Wyoming and got hired. 'Course Art wasn't very happy. But I said, "Art, you know damned well on my application I said I was a farmer, had a farming background, and I wanted to get into the feed business. I was sent to Cobourg to work all week long. That was no fun. Dad had to cover things, hire somebody. Then an opening came up in the Niagara Peninsula and that worked out great because I could live at home, work the farm, and sell feed too. So that worked out very well. I worked a couple of years for Swift's.

Marion

Born in New Liskeard, daughter to
Margery May born on an Ojibwa
reserve—
her mother dying in birth.
(Daughter to Duke, Welsh miner plundering
the underground. Dead at thirty-six, car smashed on
a black-night road.) Moved down to Hamilton
at four, words sharpened on Barton Street.
Schooled at Delta, hired by Bell, wore swish New York
clothes—listened to opera.
Met me—runner-farmer at twenty-six. A fast courtship—six
months, two rides on the Flyer at the Ex sealed the deal.
Married. Had our first son.
Then a girl. Then another girl. Years, nothing but
pabulum and diapers and bits of sleep, some laughter.
Tears. Lots of tears. Moved her to the quiet of the
farm. Not even a car horn.
Some say she gained when I moved her there—trading in
bridge for canning, shopping for hulling tomatoes. Others say
she lost. Hard to say.
This kind of life. That kind of life.
All I know is that she made it all the way to the end. And I am
thankful for that.

Marion, one year married.

My Wife

MARION CAME FROM THE CITY, DIDN'T KNOW ANY-thing about a farm, and it was quite a shock for her to realize how much work was involved and how late you'd work, how early you'd start—you'd work until dark, sometimes ten or twelve hours a day.

Then, Marion was pregnant and I was out combining. Marion was walking all morning and afternoon, and my mom kept on telling Marion that she needed to go to the hospital. Marion was stubborn, said she wasn't ready yet. I was on the combine, combining the wheat. I thought I saw something out of the corner of my eye. My mom was waving a tea cloth, and I knew I was in trouble. I took off. I got her to the hospital. A cop chased me all the way. Twenty miles to Brantford. I was speeding. When I pulled up to the hospital, he took one look at Marion and said, "Good luck," and took off. He didn't want anything to do with it.

So we lived on the farm in St. George, in a little pink wartime house that was towed in on a flatbed. Mom and Pop stayed in the house. Marion seemed to like living on the farm, except now she had kids and she was worried about the pond. I kept ducks, geese, pheasants. I built a dry-shed. She was worried that one of them was going to fall in the pond and drown, but they knew enough to stay away from the pond.

St. George was a great place. There were other ways I made money. One winter, Ontario Hydro was running a tower line, a steel tower line across the back part of our farm, and you couldn't stop them, but they would pay you something for the tower they put in. They would also pay you for the trees you'd cut down—where the lines went through—so I took the job on. I worked most of the winter.

It was all ice there. It was a swamp. So I cut all the wood down, burnt the brush, and come spring the trucks roll in with the equipment and a big tractor trailer.

We had a massive front lawn—probably an acre and a half of grass. That was just the way the house was set up. I went out and had a nice chat with the foreman. I said, "I haven't been paid yet," and he said, "Don't worry son, you'll be paid." I said, "Let me tell you this much. I worked my ass off all winter cutting this wood and I was to be paid months ago, and I haven't been paid yet. So you don't put up the steel line on my farm until you pay me."

He looked at me and he said, "You're kidding," and I said, "Nope."

He goes and gets into his cab and he phoned down to Toronto and told his boss what the problem was. The boss down there said, "Tell the kid not to worry about it—I'll have his cheque up to him the next day." He came back and said, "It's all settled. I'll have the cheque up to you the next day." I said, "Okay, tomorrow you start work. But until I have my cheque in my hand, nobody drives into my yard."

A guy came out in a Cadillac with a black hombre hat—a big-shot lawyer. "You're not exactly reasonable, sir," he said. I said, "Oh, I tell you what. If I don't pay my hydro, you cut me off. Y'know—I make an income from that wood."

I got my money.

A family pet inspects the damage at Carlisle home.

Family Left Homeless In Fire At Carlisle

CARLISLE — A Carlisle family clad only in night clothes stood in sub-zero weather early today and watched helplessly as fire destroyed the first floor of their home.

The blaze did an estimated $6,000 damage to the home of Mr. and Mrs. Barry Lush.

Mr. Lush said he was awakened in his upstairs bedroom about 6 a.m. by the smell of smoke.

"When I went downstairs the kitchen and living room were filled with smoke and flames were shooting through the roof," he said.

THE COUPLE'S three children, Curtis, 6; Laura, 4; and Barbara, 3, were hastily wrapped in heavy blankets and carried outside by their parents and Mrs. Lush's brother, Trevor Chedgey.

"I grabbed Barbara from her bed and ran outside without waiting to put my shoes on," Mrs. Lush said.

The family was later taken in by neighbors, Mr. and Mrs. Tom Gillespie.

"Another five minutes and we would have had it — I am amazed that the Waterdown and Freelton fire departments were able to contain the blaze as well as they did," Mr. Lush said today.

HE SAID the personal effects, including furniture and electrical appliances were destroyed or heavily damaged by the intense smoke and heat.

"We had just finished our Christmas shopping and all of our presents were destroyed b the fire."

The family will be stayi with Mr. Lush's brother, Pet at 4045 Apple Valley Lane, F lington, until other arran ments can be made.

Feature article in *The Hamilton Spectator*, December 19, 1963.

When Things Burn

I WAS ALWAYS AFRAID THERE WAS GOING TO BE A FIRE. I had already burned one barn down. I was raising turkeys and I was working for Swift's at the time. The turkeys were only two or three days old. One of the heaters backfired and it burned the barn down. I'm down in Beamsville when Marion called. The neighbour was there and he said your wife's really upset. He said the barn is burning, and I said which barn, and he said the one the baby turkeys are in. Will McCombe's up on the roof of the other barn stomping out the sparks. I could've lost both barns. I lost the turkeys, though. I had a mess. I had to clean up the timbers and piles of rock.

When I was working for Swift's, Pfizer came along; it was just starting up a special Ag division. So Pfizer called me and I went down to talk to them. I didn't negotiate that well. I didn't get a big pay raise like I should have, knowing what I know now, but I started, and I really enjoyed the work. It gave me independence. I was selling the St. George farm at the time, and we moved to Carlisle. The kids were really small: one, two and three years old.

That house burnt down, which was a trauma on Marion. She had a hard time getting over that. So did Curt. He was old enough to realize.

The night of the fire was close to Christmas—a week before Christmas. I didn't like being on the road because clients wanted to party and drink and if you said, "No, pass," they sometimes took it as a bit of an insult. So I decided the week before Christmas I would head down to Peterborough early and start with the Quaker Oat Company Monday morning. This was Sunday. I started to go and Joe Wheatley drove in. I had hired him for Pfizer. He drove in with his wife and his kids and they stayed for supper. There was no sense of me driving down in the dark, all the way there, so I said I'd get up early in the morning and head down.

That was the night the fire started, caused by a badly installed fireplace. Built by a guy who knew how to build them, but he didn't put any flue in and there was no insulation between the pipe going up and the brick and it overheated that night. We were very lucky to get out. Very lucky. I look back at it now and think, my God, I owe my life to those two dogs I had, Duke and Charlie. They woke me up. I couldn't smell any smoke. I came downstairs because I could hear the dogs barking, and the kitchen was ablaze. So I rushed up and said, "Marion, the house is on fire. Let's get out of here."

"What? What? What?" She ran to the hall window and opened it up. She wanted to throw the kids all out.

I said, "What are you doing there? That's the roof that is burning."

I ran and I picked up our two-year-old, Barbara—she had a little nightie on—and I ran down the stairs. The stairs went straight out to the front door. We never used the front door—we used the side porch door—and there was snow piled up. I couldn't get the door open. I really had to slam it hard. I got it open and I pitched Barbara out in the snow. I'll always remember her standing there—one foot hauled up around her waist, screaming her head off, fists clenched in the air. She was freezing to death.

I rushed upstairs and got Laura. I threw Laura out and ran back up the stairs to get Curt. I couldn't find him. He was hiding under the bed—panicking. I reached under and grabbed him by the feet and scraped his back on the metal springs. I had a hard time getting Trevor, Marion's brother, up. He was dopey as anything. I finally got him out. Then Marion and I and the kids hopped in the car. I had the car ready, the keys in it and the luggage, because I was going to take off in the morning. I drove over to our neighbour, pounded on the door, and I left the kids.

I went back and by now the house was just an inferno. A couple of firemen there said, "I think there's kids in there; we gotta go in." They were putting gas masks on. I said, "No, I got everyone out. All three kids and the wife."

They looked really happy because they weren't interested in going into that house.

———

We got burned out at Carlisle, then we moved to Kilbride, and we moved from Kilbride to Plains Road in Burlington; that's when we got rid of the horses. I sold Lucky and I ended up selling Black Jack that summer—kids in the neighbourhood sure liked him. I took Charlie, the big lab, with us. But I left Duke, the other lab, there to stay with the horses. He liked the horses. We left the horses in the barn for two months, and I'd drive up every morning and feed the horses, because I had Frank Loan's horse and the colt, and I had Lucky Strike. I went up there one morning and opened the door and the dog was gone. I never found him. I put an ad in the paper, talked to all the neighbours. I guess he went to the house, couldn't find me, started tracking and left. I always hope he didn't get killed. I hope he got picked up by another family.

Then Charlie got killed on Plains Road. I kept him chained up, but the neighbourhood kids came to play with him. He was big and gentle, and he loved to swim. You couldn't keep him out of the water. He just loved water. He got away from the chain, and he chased a rabbit across the 403 and got hit by a car. I'd had him for about three years. Beautiful purebred.

The house on Plains Road was a beautiful home—it really was. It had a tennis court, a garage, a carport. Mom really liked it. Right after that, I got a promotion to sales manager for Pfizer, regional manager of Ontario, which increased my money. And I was lucky—two of the three years, I had the top sales force. Really nice guys.

I had fired the guys who weren't working out. They said, "Why are you the boss? You don't have a college education." They were the poorest salesmen, and they felt that the company owed them a living. Entitled. So I got rid of most of them and I hired my own guys. They were great. They were all different. Joe was a hard-working guy, an Englishman. He knew farming pretty well because he had farmed in England. Joe Wheatley. I hired Johnny Thompson, because I had worked with John at Swift's.

Then I quit Pfizer because I was going to have to move to Montreal. So I quit and worked for my brother. My boss was thunderstruck. To get anywhere with Pfizer, you had to move to Montreal, and I didn't want to move to Montreal.

My father and his friend sitting on the front porch steps.

Owen Sound: 1976

I went back to farming because
I couldn't stand wearing a suit and tie anymore. I
couldn't stand the cars and the sidewalks, cold.
I couldn't stand the always shine of my black leather shoes, the
weight of my duffel bag,
the endless shuffle of paper, the in
and out box, the telephone calls.
That was no way to live.

I went back to farming because
I had to. I had to get back to the land. The
sun. The earth. The barns.
I had to get back to something before all this.

Are you crazy, everyone said?
Oil prices and inflation skyrocketing. Worst
possible time to buy.
Worst possible time to buy a farm.

The Banks. Sonsabitches.

Before interest rates rose, they said borrow money.
After interest rates rose, they said borrow more money.
Many fine men hung from barn rafters
because of what those banks said.
Many fine men drank themselves into the ground
because of what those banks did.

So, I looked everywhere. All around southern Ontario—
Belleville, Orillia, Peterborough—until I found that 148-
acreage just west of Owen Sound.
Until I found those two barns, empty and waiting. And I filled
them.

Flossie with her piglets.

The Last Farm

WHEN I CAME UP TO OWEN SOUND, WE CAME AT ONE of the worst times in farming in thirty years, when taxes were up, when interest rates were up. You had to borrow money to farm—eighteen and twenty percent. Couldn't believe it. Farms were being foreclosed. Banks were notorious for that. And you had to borrow. You couldn't buy cattle for cash. You were trying to build up your equity.

The kids would get up and help feed the pigs in the summer. Mash was barley. Concentrated soybeans. Mixed it with water. All three kids would feed them, mostly on the weekends. They couldn't go into a pig-barn and go to school. I had these two special pigs, Skippy and Art. They had a weakness. Skippy got rolled on when she was a baby, so her back legs were paralyzed. So I took her out of the pen and made her a harness. Secured it with a rope that I fastened to a pulley so she could move around freely in the pen. She loved it. You could go right up to her and scratch her under the chin. She was that friendly. And Art—he got attacked by the other pigs. Really carved him up. Pigs can get pretty nasty with each other. So I took Art out of the pen and let him roam freely with Skippy. He never left her side. They were quite a pair, and when Skippy eventually died, Art sunk into a real depression. You could tell he was affected. He wouldn't eat. Just stood there. So I put Art back in with the other pigs, but they started attacking him again, so I had no choice but to let him just roam by himself outside the pens.

    He was like a dog, that pig.

Summertime, I'd let my sows go out into the field, two or three at a time. Once they got used to each other, there was no more fighting. I had a cow break her leg. She was running and she tripped into a groundhog hole. I had to shoot her. But you treat your animals gently, you treat them well. I lost my temper a couple of times. One sow jumped out of the hog barn and she went into the pond and had the babies in the pond and they all drowned. That's a loss. You have to feed them. I was averaging about twenty pigs, twenty wieners per sow. That's high. But I really looked after them. I'd stay up at night, even in the cold winter, when they were farrowing. Every once in a while, they'd panic, and they'd step on them. And they'd step on another one. They weren't doing it on purpose. I loved animals so much that I'd get a great kick out of it. Best job I ever had.

I had a lot of good pigs, but I never had a pig as good as Flossie. When I started, my son, Curt, had to go out looking for her. I came home Wednesday night and she still wasn't back, so I went out in the field and she was out along the fence line. She found a bale of straw and tore it all apart and had her pigs out there. Nine babies. She wanted privacy. She did that twice. Next time, she went out to the bush. She dug a big hole beside this maple log on the ground. We looked and looked and looked. We went out to the bush and didn't find her. Marion saw her ears poking out from behind the log, so we left her out there for three days. We could see her coming across the field with nine little wieners.

Two sows and a young boar.

Piggery

I HAVE A PICTURE OF THREE PIGS. THEY WERE JUST OUT in the barn yard. I pastured all of them out in the fields in the summertime rather than in enclosed pens, which made them a lot healthier because they liked rooting around. These are two sows and a boar. A young boar. I'd mix up some chop and some concentrate and some water and mix it into a porridge and pour it into a trough and they ended up loving it. They cleaned it up, they really cleaned it up. They knew when feeding time was because they'd start to grunt.

Gosling and geese.

Geese photographed from the barn roof.

The Goose Man

ONE YEAR WE KEPT THIRTEEN HUNDRED GEESE, DAY-old goslings right through to market for Swift's. Out of the bunch, I kept maybe a dozen. They hatched every year, and we had geese for years to come. I also kept a lot of turkeys, and we'd butcher them every fall and sell them off at Christmas time. Sometimes, I'd climb up on the barn roof. It was easier to take a picture of the birds from up high.

Thirteen hundred goslings. The biggest problem was foxes and coons, 'cause you couldn't keep the goslings in the chicken pen for too long, they had to get out and graze, two-thirds of their feed was grass and one-third grain. They'd be out in the field and coons would come by and grab them at night. I had a station wagon at that time and I put a mattress in the back, and I slept out in the station wagon with the back pointed toward the field with my gun. I shot a couple of coons and one fox. It seemed to stop them.

Geese are an intelligent bird in some ways and stupid in others. I lost about eight geese that fell in a hole and they just piled on top of each other, so I had to go around and make sure all the groundhog holes were filled. In a farm where you're bringing in hay, you have a lot of groundhog holes. It was just a waste of birds 'cause they were a fair size, and they'd fall in the hole and sit on top of each other and squash themselves.

Probably 1,150 went to Swift's. I lost about 150 altogether. Swift's sent out great big trucks with cages and five men and they gathered the geese up. My youngest daughter, Barbara, was really mad because she thought they treated them really rough. But it was their job to deliver them and take them away. My job was to look after them. For that I got paid a certain amount of money.

We put them in feedbags so they couldn't break their wings because they'd be flapping away. That was when we were taking them to the processing plant in Durham. There was a good profit in geese.

When I sold the farm, I still had three geese left. Pets. They were probably twelve years old. They were watch dogs. They kept all the garter snakes away from the house. They loved snakes. One would grab one end, and another would grab the other end, and they'd run around the house chasing each other with these garter snakes. Marion didn't like the snakes, so we hung onto the geese for that reason.

They're good, clean pets. The last three years, I never even put them away in the winter time. They were quite capable of sitting in a snowdrift as long as they got grain.

Feature article in *The Sun Times*, December 17, 1983.

Christmas Delivery

THE STORMS WERE REALLY BAD AT THAT TIME—THE late seventies. Sometimes you couldn't get down the lane, so I drove my car down to the edge of the laneway, lugged the geese down in bags because they'd already been processed, and people would come and get them. I'd have the names of whoever had bought them—Mrs. Walker, Mrs. Cross, et cetera.

When I sold the farm in 2010, I gave away the last three geese I had to a farm on the way to Wiarton. I couldn't put them down. There's no guarantees what will happen to them. I had those geese a long time. They still tried to hatch eggs, time and time again. As the geese got older, their eggs became infertile, but they'd sit on twelve eggs and sit on them and sit on them until the eggs would blow up. They were rotten. Full of gas. Then they'd hang their heads and trot off, and I'd have to take a pitchfork and clean their nests. They made their own nests along the fence line or somewhere safe and hidden.

They were great birds to look after. They knew me. I know when I had the thirteen hundred, they had the run of the whole farm and they'd go back to the bush and eat there. I climbed up on the barn roof one day and took a picture of them. I could see their trail going right back to the bush, and a guy from the local paper came out and took some pictures and he was astounded.

The last five geese all ran to the barn one night. They were hiding in the barn from predators. Two of them got it though. Coons. I also had some trouble with the neighbours' dogs.

Two of the last three geese bathing in the pig trough.

Dad captures the last goose to go to the Bumstead Homestead.

Geese

Those geese flapped, broke wings,
fought, snapped snakes, hatched goslings—sat on eggs
until they exploded.
    Sat quiet and white
on snowbanks—fell to coons, fox, neighbours' dogs. Sold for
Christmas geese—now that was me thinking what can I do
now for a few bucks?
    Followed the trail of seed in the barnyard—pecked at
our scraps, bathed in pig troughs,
fled to the barn for safety—later two carcasses,
    dead.
    Two out of five, now three. Oldest
geese in Grey County. Twelve maybe fifteen years old. I took
those last three on that last day, rustled them into a feed sack,
feathers scattering, helluva noise
    and drove them that last time on that last day to the
Bumstead's on the county line,
    where—hard to say—dunno—what
    they're doing now.

Farm fresh eggs lined up on top of the dryer.

Eggs

I head to the chicken pen first thing in the morning.
There will be squawking, a commotion
of feathers. Smell! Pinch your nostrils tight.
    But once they know it's me—they'll settle down. Flap just a bit.
Then I go over to each laying box, see what they've left me.
One, two, maybe three eggs. They are beautiful and brown-oval. I say to
myself, My God, they're perfect. Then I put
them in my wire basket,
a couple of dozen at a time
and bring them into the summer kitchen.
    I bring in the smell of chicken and straw and manure, as well.
    (Marion keeps me out of her kitchen.
The summer kitchen is for me and my washing and cleaning.) Then I run
them under the tap, one at a time,
turn them like you'd turn a wrapped present.
Then I line them up on the dryer and
just look a while.

One of the chickens peeking out of the chicken pen door.

Wheeling a load of manure out of the pig barn.

Cleaning

EVERY DAY I'D CLEAN THE STALLS. YOU COULDN'T LET the manure pile up. I'd use the manure on the land. I really liked a clean barn. I'd put shavings down. They absorbed well. Then I'd put straw on top of it and it would work very well. The Owen Sound barn was a cattle barn, but I turned it into a pig barn by taking out the stalls and putting in separate pens and the sows could have their litter in that pen all by themselves. But it was a labour-intensive deal. I had to really go through the barn each day with a wheelbarrow and gather up everything. I'd work on two or three pens a day, so over the week, all the pens got cleaned up. I had a chap from down near Listowel come up every year, and he'd whitewash the whole interior of the barn, as well as the outside on the stonework. It really just freshened everything up; it killed the flies and the smell.

He'd spray in the barn with the sprayer. He'd spray with lime. We'd put an insecticide in it as well and it worked really well. It made it smell clean when you walked in—and it was bright. You could take city people in there. You take pride in your barn. I had two good barns. I played music in the barn—always had music so the pigs wouldn't get startled by a bad noise. The music seemed to quiet them down. It was a constant thing. Like white noise.

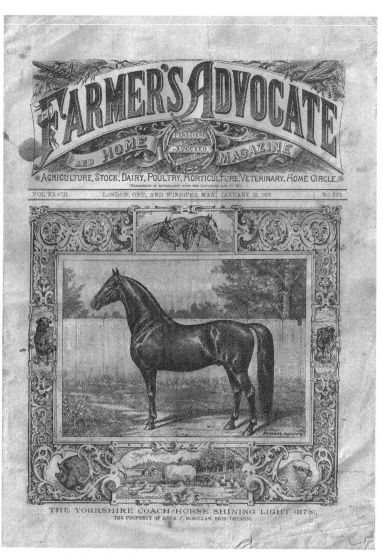

The Farmer's Advocate and Home Magazine. January 15, 1893.

Looking Back

YOU THINK THE PROBLEMS YOU HAVE TODAY ARE NEW problems, but they aren't. Not at all. I bought a bunch of old *Farmer's Advocate* newspapers once—five of them at a farm auction down near Chatham. It was interesting because you read the problems they were having back in those days, before motorized vehicles came into play. So there was a lot written about horses and special cows. I'd always want to see what animals were selling for in those days compared to the seventies. There was real estate in there—one of the finest farms in Manitoba was going for twenty dollars per acre. It showed how values had changed so much from that time. I liked reading the advertising in terms of what people were putting in the ads to sell—lambs—what they were asking for. This was the forerunner of the big agricultural magazine. It had all the information that a farmer would need—how to overcome weeds, how to overcome the present agricultural depression. It's a history book to me; it showed how it was many, many years before.

They had a whole veterinary section: Catarrh and sheep. You had to be a veterinarian. A section: Why the Boys and Girls leave the farm. That still happens. The problems two hundred years ago are the same as today. Listen to this:

> *How should we keep them [boys and girls] at home? In my mind, the solution is easy. Give them, in as great a measure as possible, the coveted pleasures that lure them to town. What are they? Wealth is not one.... Let the boys have some stock...it will yield them something better than money. A harvest of knowledge. (Carson, 1893, pp. 35)*

A lot of the problems that were shown still exist today. They existed in the fifties. Young people were trying to leave the farm. They didn't see the value in it until they were older. There were a lot of jobs that weren't hard work—they could get a job in the store. They could see the bright lights of the big city. They could make a lot more money driving taxi. You had to like farming. You had to like growing things, raising things. And you had to like the family you were working with—your own family.

There was quite a lot of spite at times between brothers, one getting more favours. It was difficult. Usually the oldest son got the farm, and the other sons went off to the city.

Pop only did a little bit of farming toward the end. He didn't grow things; he took care of things when they were ready to come down. He liked driving the tractors, but he was scared of the tractors at the start. You're talking about a man who never drove a car until he was in his forties. He had to learn. But he liked being on the tractor. Eventually, he got a big kick out of it. I have memories of him on the tractor at dusk, a cigarette—a Peter Jackson—in his mouth, and him wearing a checkered flannel long-sleeved shirt.

Owen Sound II

WHEN I CAME TO OWEN SOUND, I WAS IN GOOD SHAPE. I was looking forward to getting back to farming after being away for twelve years. But we came back to farming at the worst time in thirty years. Interest on the money was so high. But you had to borrow money to farm—at eighteen to twenty percent. I couldn't believe it. Farms were being foreclosed left, right and centre. Banks were notorious for that. But you had to borrow cash to farm. You were trying to build up your equity. Everyone in the family pitched in, especially those first couple of summers. Both Laura and Barb would help feed in the summer. Mash was barley and oats chopped up and mixed with soybean meal, and then it was mixed with water to make a slop. We fed them in troughs, the most economical way to raise pigs to two hundred and ten pounds.

Drunk Piggery

I HAD SOME VERY NICE SOWS AND THEY WERE ALL coming in at various times so I didn't have all the pigs at once. I liked to have at least one sow a week that was farrowing. To do that, you had to have your own boar. So I bought this boar down country and named him Joey. He sure grew fast. He knew what he was there for, and he was in such good shape that all the neighbours wanted him. So I'd trot him down the road with a stick and he'd just march along. Marion thought he was vulgar. I don't know what she thought—how these pigs were going to be born. The sows were a good business. They provided money coming in each month year round. When I left it for twelve years and came back, the hog business took a dive. It was down for five or six years. I worked my butt off and lost money on each sow.

I had another sow—one of the biggest ones I had, probably five years old. She walked down in-between the cattle stalls and got her head into the ensilage in the silo. The ensilage was just put in, so the moisture was leaking through the bottom, and it was alcohol, and she got drunk. She drank and drank and drank, and she got so drunk she couldn't move. There was just a narrow walkway that you could push a wheelbarrow through to get the ensilage out of the tower silo. I had to leave her there. She stayed drunk for a week, until the ensilage stopped weeping. Once it dried up, she finally sobered up. She staggered backwards—you'd never have lifted her up as she was close to four hundred and fifty pounds. I thought I was going to have to put her down. She was determined to drink that stuff—she loved it.

I remember going in every morning to see if she had sobered up—the ensilage was corn liquor, whisky. The ensilage was in the silo. You put it in green and quite wet and it would dry out in the silo. The moisture would drop down and come out the bottom. The sows were set up so the water would drop out and run into the laneway. Except this damned sow got out of her pen, and she smelled it and went up, and you couldn't turn her around, so she kept drinking and got drunk and stayed drunk for a week.

My son, Jack, feeding the grey mare.

Fence

He'd get up on that cedar fence, the one
I built when I first got here and just look.
His eyes fixed on the
grey mare I had in the barnyard. And he
would wait—sugar cube in hand— until the
mare would amble over.
Nostrils flaring, tail swatting flies. That was
all he needed—
a horse, a horse licking a cube, small and
sugary, off his small hand.

My father fixing the roof of the barn.

Barn

She hated it when I spent any money on it. When I fixed
the roof. Christ was she mad then.
Tied a rope around my waist and got up there anyway.
I nailed all those sheets of metal. I wasn't afraid of falling. It looked
different up high.
I could see the turkeys—the reds of their necks. The
pheasants pecking the ground.
She always made him stand at the bottom by the truck, though—where I
had tied the rope. She made him watch for me. Made him look up. I didn't
ask my son to climb up on that roof. That was my job.

The House

He never fixed the broken doors. There were no locks
the thirty-five years he lived there. Windows icicled in the cold. Floors
creaked.
The fridge bumped Freon at night.
    But the barn. He'd throw a thick rope around his
waist—tie one end to his pick-up truck, and climb that gilted
ladder—climb that beanstalk to the top.
No giant waiting. Just shingles, old and fodden,
Ready to be ripped. They fell like flotsam to the ground, scattering the
geese and turkeys. Then he shined up the roof with new tin sheets. God,
you shoulda seen it then.
You could've skated on that roof.

Dad's Garden.

He watered flowers, he watered all living things.

Return to the Garden

I'VE BEEN PLANTING A GARDEN SINCE I WAS SIX YEARS ago. Pop wasn't around most of the time; he was in the hospital from gas attacks in the First World War. He'd be out in the summer but he couldn't dig the garden very well.

The first garden was on Guelph Line in Burlington. Pop had a little chicken farm. He rented the place, and we had a garden there. I'd plant all the greens and carrots and potatoes. My mother helped me. She was a really good gardener. The last year we were at Guelph Line, Pop bought a scuffler. It was a machine that went down between the rows and took the weeds out. But he couldn't push it himself. So he tied a rope around me and tied it around the scuffler, and I'd walk ahead and pull it.

Growing things was something I learned early in life. Torrance Street in Burlington is where I really took off with gardening. I grew blueberries, strawberries—I grew enough so I could ship them off to Toronto.

I had a big garden on Brant Street in Burlington, a whopping garden—six acres of tomatoes. One acre of strawberries on Walkers Line. It took a lot of work, but it was a great sense of accomplishment. When I moved to St. George, I was so busy with the cattle, that my mother did most of the garden. I'd plow it, and she'd take the scuffler.

When we moved to Owen Sound, then I had a really big garden—raspberries, strawberries, potatoes, tomatoes, beets, carrots—the whole gamut. I never had a disaster with any of the gardens. They all grew, some more than others. There was no such thing as a drought.

Planting: May Long Weekend, 2012

I'M GOING TO GET A ROPE. YOU FOLLOW ME. WE'RE going to put in tomatoes, peas, beans, radishes, all the regular stuff. The only thing I don't have is lettuce. I should've picked it up.

I didn't have a chance to do a garden last year. I was feeling too rough. The garden is going to make me feel terrific. Animals and gardens are different all together. Animals need attention every day; a garden doesn't.

We'll make rows now. We'll stick in a broken stick and tie a piece of string. We'll start with a few seeds in each hole. We're going to make up lines here. Mom used to help me plant the six acres. I'd get it nice and cultivated. We'd walk with boards on our feet and nails and mark the garden like a checkerboard. We'd walk and mark it.

We'll put the cages around the tomatoes when they get bigger. I'm worried about coons and rabbits getting at the tomatoes. We need two feet between the tomatoes—you got to remember that they like to spread out, tomatoes.

Jessie and Jodi in the truck

Truck

All I had to say was "truck," and they'd come running.
Those two cross-hounds I bought as a pair, couldn't/wouldn't separate.
Marion took one look at the sister, and she said, "We'll take her, too." So
we did. Jessie and Jodi, Curt
named them—one more Shepherd, one more hound. Jessie, the
Shepherd didn't like rain or storms, would crawl belly-to-the-floor
into the safety of the bathroom. Jodi would whimper and lie close to
my feet.
But the look on their faces when they knew they'd
be going for a ride in the '78 pickup. Around the barnyard to unload
feed, or to pick up wood. It didn't matter. I'd open
the door and they'd just jump in, Jodi on one side, and Jessie on the
other. Tongues long and sloppy.
The day I drove into town to pick up feed, my right arm suddenly stopped.
The dogs were in the truck. I couldn't just go to the hospital. I had my dogs.
So I asked the guy to reach for
the keys in my pocket, and I drove home, my left hand on the wheel—Jessie
and Jodi licking at my limp-stroke-frozen arm.

The hay mow.

Haying

When you go inside, there are all sorts of

things. Old grain, horse dander.

The hay mow leans a bit.

'Member when you kids lifted the bales off the bailer,

screamed when you saw a snake.

disappeared between the haystacks,

your arms saving you at the last minute?

They were times all right. We were all together. That's

what I remember the most.

SECTION II

I didn't leave the farm—I still farmed in my mind.

Changes

THEY DON'T BRING IN LOOSE HAY ANYMORE. THEY don't bring in small bales anymore. They have big round bales, and they leave them out under plastic. They don't keep their cattle in a barn anymore. They keep them in sheds. Farmers used to have a lot of kids or they could hire help for labour and board for a couple of bucks a week. But you can't do that anymore. Farmers got into bigger farms. They bought bigger equipment and bigger tractors. They couldn't drive these tractors into a bank barn. They'd fall right through the floor. Some of these tractors are worth $150,000. So they changed to these big round bales and they'd stack them outside and put tarps over them.

All of a sudden, barns became specialized.

In my day—the barn in St. George is a perfect example—there was a cattle section, a pig section, a chicken section. In those days, the farm was family-operated, you didn't specialize in one thing. Every farm had pigs. Every farm had chickens. Every farm had some cattle. Every farm had a market garden. They don't do that anymore. The barn is part of the farm, so they're not going to tear it down if they've kept it up. Even if they don't have livestock, they need to put their equipment in there. They can store their small equipment in there—boats, any number of things.

Other times, they don't have any use for barns. That's when they let their barns go. When they let their barns go, they tear them down. They don't want to pay for the upkeep, because any building requires upkeep. You gotta paint them. You gotta fix windows, doors, or they fall apart.

You drive by some farms where they've never used the barn in years, and it's almost ready to collapse. There were three of them along 26 highway to Collingwood. Each year I'd drive by, and they'd be in worse shape, and one day, the wind just blew them down. It used its time. It's just a pile of wood. So people come around and cut it up for firewood. They crumble onto the floor of the wood stove in a heap of grey-white ash.

Farming has changed. People aren't building big barns anymore. They're building sheds! They're building sheds because the haying has changed. You don't bring in loose hay. And you don't bring in small bails anymore. Me. I'd go up to the top of that barn. I'd climb that wooden ladder. Get up there, throw down six bales, and they'd come crashing down onto the floor, and I'd come down and break them loose with a pitchfork. It was hard work, but it was satisfying work.

My father's sailboat in the bottom of one of the barns.

The Last Barns

1976 to 2011.

That's the longest time.

The longest stretch. No one was going

to make me sell my farm until I was ready.

I still wasn't ready when it came to the time.

The one barn was good.

The other one not so good.

The doors didn't slide so well. The geese would go in there and hide from

the coons. I lost two of my geese

that summer to the coons. The other three hid in that

barn. A couple fell through the floor

and landed in the bottom where I kept my boats.

    Safe down there.

Keys

In thirty-five years, we never had any locks
on the doors. *Hell, didn't have to.*
    I had my neighbours.
    I had my geese.
    I had my two dogs.
But then, I had to sell the farm. Had
to have a locksmith come in a big
black truck. The words Lock Ready.
He fitted each door. The front, the
side, the back with a lock, the way
you'd fit a prisoner with an ankle
bracelet.
And then he shut those doors.
And nobody ever came through those
doors again without knocking first.

Trees

When we first moved up there,
I lined the driveway with hockey sticks, black tape
still cloyed around the handle.
Then I planted trees—12 feet apart, 24 on each
side. They grew each year
with the hay, with the kids, with the highs and falls of
the weather.
    They grew until
there was a small forest straight
as an arrow to the blue rectory church door of the
farmhouse.
    When we left, they cut down each tree. I
could hear each tree-body fall.
It was a hard and cold falling. The measure of our thirty-five
years there. They wanted the space and wind
instead.

My father walking into the empty barn, except for two TVs.

Selling the Farm

I didn't want to sell the farm. I
could've stayed there.
I could've stayed there until the end.
But my grandson wasn't going to grow up on this farm. My son was
no longer with us—no—*he* was a great boy. My two daughters.
Well, they do what they do.
    So I sold the farm. It was that easy.
It was that hard. And I watched my kids
haul everything out of those barns—my feedbags, harnesses,
ten-inch nails, old windows, ropes, electrical supplies, skeets.
    I watched them put everything
in a pile, then I'd walk around that pile with a stick, check the
pile, poke a bit, put a few things
here and there.
    Mostly back in the barn where
they belonged. That would get them upset. They'd haul it
out again in the dark and I'd check the pile in the morning,
and put it aside.
Eventually, they won.
    See, we both won.
I brought my pile to the new house, and they cleaned
the barn. Cleaned it good like you do before you're ready for
something. Wash up good. Christ—
    you could have eaten off those floors.
And I walked around in that hollow shell—hallow—
    Looked up at the rafters—those hay/straw-barren rafters—and just
    looked.
I couldn't believe it. The emptiness. And I thought, There's
no way in hell hay will ever fill these barns again. And I was
right. They took the barns
down right after that. Right to the bare foundation.

And I will never go back.
    —Hallow—
    —Hallow be *thy name*—

My father gazing at his old treasures taken out of the barn.

The pile of scrap wood taken from inside of the barn

Inside of the barn, empty.

Fallen debris, including a For Sale sign. Lush Realty is owned by my uncle, Peter.

Song

My pitchfork is raised.

I fall onto the prongs of those soil

cutters. No one looks.

Reflection

In the morning, I looked out at the fields. In the evening, the fields looked out at me.
Mornings always made room for another morning.

When I farmed, I lived, when I farmed, I felt, and when the farm left me, I didn't leave the farm.

I still farmed in my mind.

Workers start dismantling the barn.

The Razing

When we got there, they had already started.

I could hear the sound of barn boards torn

from their hewed sockets. I could hear the yellow voices of old

straw. My daughter went first. She led me through the long

tired grass of the barnyard.

I could hear the sledgehammers. The joy

of their pounding. Taking away, breaking down. They.

It wasn't their barn. They weren't a part of it.

They had no feel for the farrowing. No memory of hay. They had

their shirts off. Like this was some sport.

My daughter kept looking up at them—*searching*.

They shouted to each other above the hammering. A cigarette fell.

And then my barn just receded into the earth, until the bareness of its

foundation—its hulking fossil of white limestone—grounded and rubble.

My father walks past the barn as it is dismantled.

Taking the Barns Down

YOU KNOW HOW MUCH THE NEW OWNERS GOT FROM those guys who tore down those barns? Two thousand dollars. Two thousand dollars for two massive, thirty-six hundred square-foot, bank barns with hand-cut wood from the bush. Those beams ripped out of my barns—they're going into furniture, they're going into slabs for a wall in a rec room, they're going into somebody's new home in the city. They're not in a barn any longer. They're somewhere else. They're somewhere else now. So when they took down my barns, I couldn't believe it. They were beautiful barns. My cattle lived there—fifty head. My sows—Flossie, Miss Pickie, Joey the Boar.

Marion—she even loved those sows. She would go out into the field looking for Flossie. Flossie always had her piglets in the field. She was different. A bit of a loner. Then Marion would go back and look for her, see her pink ears sticking through the long grasses, and she'd yell, "Barry, here she is, I've found her," and she'd be so damned relieved, worrying that those babies—Flossie's babies—were going to die. I'd just laugh and say, "Flossie's not going to starve her babies, Marion." No animal—no pig, no mother—is going to purposely starve their babies.

That was a hard day, seeing the barns taken down. I couldn't wait to get out of there. I had to go. My boats were in there. My motor boat. The one I sunk off the Muskoka docks. Remember? And my sail boat. I built that boat myself when I was sixteen. Sailed Vic Damone around in that boat in the Burlington Bay when I worked at the Brant Inn. I had to get my boats out of the barn. Those guys didn't care about my boats. They just wanted the barns down. I could hear the hammering, the beams crashing.

The barn cats. What was going to happen to them? I had made a deal with the new owner. She was supposed to look after the barn cats. I told her she'd need a few cats in there. Every barn has cats. She promised me she'd feed them. I'd go out there every morning and leave food and milk. Now what was going to happen to those cats? Those guys with their shirts off, tearing my barn apart, they don't care about my cats. They just want the barn down.

The demolished barn, photographed from the concession line.

The Barn, Broken

The beams. The posts. The lofts.

The columns.

The bays that separated the columns. The rafters.

The purlins supporting the rafters. This is the barn they tore down.

Victorian concrete ornaments that topped the gate posts at the laneway entrance.

The Newels, Lopped

    They were just lying there in
the grass—cold, weighted.
And I thought to myself—
My God, they really did it. They took the barns
down. They took the farm apart. They took
the balls right off the farm.

The house and the fallen barns.

Aftermath

They yanked and they pulled, they cut

and they sawed,

they threw and they heaved until every piece,

every beam, every barn board, every sound and every colour of those

two barns lay heaped in a mess of bruised wood.

Then they put down their hammers and pickaxes, they

put away their saws and their wire clippers, got into their

big fendered trucks and drove off—

these eight bare-breasted men—had *taken down my barns,*

my bit of history, had scraped a bit of the landscape away.

I began to feel cold, I couldn't breathe.

    I sank into that grass, my body flattened down, too. My heart.

And I thought, what am I to do now?

Where am I to go? And I didn't know the answer. So I got

up and just kept walking.

Dad as a teen with one of his first dogs, Bud.

A Man and a Dog

HE WEIGHS FIVE POUNDS SIX OUNCES, TINY HANDS and feet, a mouth that can't latch onto his mother's breast. She feeds him formula and special vitamins from a dropper and grows him to five-foot-six. He wants to play football, but he's too small, wants to become a fighter pilot, but he's too young. He runs instead. He runs every day in army boots, a white peaked cap, red shorts and a maple leaf tank top. He runs from St. George to Brantford, from Hamilton to Toronto. He runs around the bay in Burlington, he runs the hills of Boston and New York, the heat of Mexico City, the rain and fog of Vancouver. He collects trophies and ribbons, lunches with the Duke of Edinburgh. He dates girls taller than him, richer than him. He courts well—always opens the door, pays for dinner and brings flowers. He spends all his food vouchers on a Danish girl he meets on the train to the British Empire Games in Vancouver.

He talks about the first time he met his wife, how he took her on the roller coaster at the exhibition—not once but twice—how she threw up in his white-peaked cap, *how she knocked his socks off*—her New York clothes—how lucky he was, this farm boy from St. George.

He says, *Well, I'll be, son-of-a-gun,* plays the ukulele, the gut bucket, sings the first eight bars of Tommy Dorsey's "I Can't Get Started" every Valentine's Day. He names his children with care—Barbara after his sister, Laura after the song, Curt after a southern boy from South Carolina who gave him his dog tags so he could feel for himself the weight of the Second World War around his neck.

He can parse plants, differentiate a good bale of hay from a bad one, oats from wheat. He loves spring and summer, winter and fall. He packs up his wife and kids every year and takes them to campsites in northern Ontario. He shows them how to bait a hook, cast a line, make the lure dance under water—when to throw back a fish. He takes them to PEI, Expo—finds a tent blowing down a street in Lethbridge. He sinks his boat off a Muskoka dock, water skis into an island, lands on a cow when skydiving.

He spends hours in his barns, inhales the sweet-thick aroma of silage, runs his hands up and down the roughened hand-hewn beams, shoots rats buried in feedbags, watches the steamy breath of his cows huddled in the lower barn, the whip and run of his horses across the field, walks the concession fields, pauses so he can watch the rhythmic line of his barns rising out of the landscape.

He raises birds—turkeys, pheasants, geese that run towards the sound of his boots in the barnyard. He guards their eggs hidden in the long grasses, sleeps outside at night with a gun tucked under his arm, shoots at anything that moves.

He watches sports every night in a black La-Z-Boy, fully reclined. He sits with the remote in his hand and switches back and forth between hockey and boxing, football and baseball. He pulls his La-Z-Boy close up to the TV, presses his ear against the screen so he can hear the sound of a hockey puck zinging across the ice, jumps off his chair when a football cleaves into the hands of a running back, shouts *Touchdown!*

He keeps his pockets full. Lifesavers, sticks of gum, broken toothpicks, a set of nail clippers, old Lotto 6/49 tickets with the numbers circled in pen, loose change, balled-up gum wrappers. He can never find his glasses, his wallet, his car keys. He drives off with piles of laundry on the hood of his car, forgets his seven-year-old son in a gas station washroom, drives back half an hour later and finds him balled up and crying on the bathroom floor. He always remembers birthdays and anniversaries, the exact time when the tufted heads of his kids crested his wife's birth canal. He can tell you when an egg will hatch, when a sow is sickly, when it's time to walk away from a crop of tomatoes.

    He knows how to build a fence line, clean a gun, remove porcupine quills from the snout of a dog. He sits you down on Remembrance Day, tells you about Gallipoli, Dieppe, how his own father—a bullet in his chest—ripped off his gas mask, sucked in the mustard gas, burned his lungs to shreds. He can sail a sixty-foot boat into the dark maw of the Caribbean Sea, tie a tourniquet, break up a field with a discer. He is agnostic, anti-Church, sings the last three lines of "The Seafarer"—*Amen, Amen, Amen.*

He loves the open. An open window, an open field, the open arms of his children, the openness of Canada. He believes in freedom, democracy, capital punishment. He watches action movies, Westerns—anything with Clint Eastwood and Charles Bronson. He intervenes in fights, gets black eyes, broken ribs. He likes money, but not saving it—prefers the pull of a slot machine, the slap of cards in a blackjack game. He buys three of everything: three weed wackers, three hammers, three lawnmowers. He buys cars, forgets to change the oil, waits till the last minute to signal left, stops to pick up hitchhikers. He talks to anybody, doesn't drink, doesn't smoke. He goes to the grocery store, comes home with sticky buns and butter tarts, *three of everything*.

He names his farm animals, splints their broken legs, constructs elaborate hoists and zip lines for pigs crushed and hobbled. He spends hours rubbing their bellies and ears, mutters *goddamned-son-of-a-bitch pigs* when he takes them to the abattoir. He always owns a dog. Dogs follow him everywhere—behind his tractor, into the barn, grocery stores, belly-crawling up the aisles.

He flies the Union Jack on July first, votes Conservative, extols the dangers of communism when his son writes to Russian pen pals. He argues with him about everything—food, music, the way the world carves itself up. His son orates human rights, the hypocrisy of American policy, corporate greed. He says, *Hey son, wanna play catch? Shoot a puck?* Not today, Dad, his son says.

He packs away the hockey sticks, baseball gloves, studies his son instead. Learns about his world, the *other world* sends his spare change to Nicaragua, spends whole summers with his son jumping the waves of Sauble Beach, watching the same movies, talking about history—long past arguing—his world just a little more open. But slams shut when he learns that cold October morning that his son, his oldest child, Curt, named after Curt Ayers, that southern boy from South Carolina, is now gone.

Gets old, gets sick, his heart falters. He talks to his dead son about what's wrong with American policy, burns the Union Jack, votes NPD, and when his dog can no longer eat, lays his eighty-year-old body down on the cold basement floor beside him, *You can't go, too. Not you, too,* he rubs his dog's ears, *There there boy,* when the dog has a seizure, *Easy boy,* as he walks him around and around the yard until even he—this five-pound-six-ounce baby now grown into an old man, this runner, this farmer, this fighter, this father who lost his son, his oldest child named after that southern boy from South Carolina who gave him his dog tags—cannot save his dog, the dog who followed him behind his tractor, and the dog just drops and stiffens on the grass, and the man does not move, just sits for a time under the willow tree in the hot August afternoon and thinks about his dog.

Dad as a toddler.

The barn, stripped of boards, still standing erect.

How It Carriages

It stands—bare, unadorned, foundation girthed to
the rough stubble of earth.
Like him, every year, it ages. Boards hooked by
wind. Lime walls
crumbled. Like him, it does not complain. It is
filled with what it needs to be with— sustenance,
life. The egregiously rickety of once-good parts
now broken.
Like him, when it's time—or when it's not time—
it will be taken
down. And it will fall and the earth will harden a
little.

The silo flush to the side of the barn.

The Barn Speaks: II

I remember hay, straw against my beams.
I remember the chyick of the grain elevator, the burn
of ropes on my beams. No man
hung from my beams. No man died of my wood. A pony
sought me out once. Old and teeth-rot,
it just lay down. His owner, that man, that farmer, that
man who let the hammer-men take me down, he sat
down beside that pony and cried.

Cat in the empty barn.

Last Words

I can't tell you how I feel about my barn. It's a
place I've always known.
Every farmer has a barn.
To take it down?
To take it down?
That'd be a terrible thing.
I got cats in those barns. *My hay.*
Thirty-five years I've had those barns. Listen a
minute. Can you hear the boards? Gnawing
of oats? The smells?
Like a tractor just getting going.
        The land turning over.
The sow farrowing in the sluice.

I had a pony not too long ago. The
pony went in there to die.
How would that pony
have died if not for this barn?

Acknowledgements

I WOULD LIKE TO GIVE SPECIAL THANKS TO MY FATHER, Barry Lush, for living and telling these stories.

Swing Beam: My Father's Story of Life on the Farm and the Barns He Loved And Lost was originally written as an arts-informed life history thesis for my Master of Arts in Curriculum, Learning, and Teaching at the Ontario Institute for Studies and Technology (University of Toronto) in 2013. Some of the text has been changed or left out.

In addition, excerpts of the original thesis have been previously published in Kwai Li's anthology *That's What They Said: Oral History* (Life Rattle Press, 2013) and in *Fishflies and Other Stories* (Life Rattle Press, 2013).

Some of these photos were taken by my late brother, Curt Lush.

Thank you to Guy Allen, Laurie Kallis and John Dunford. All of you are in this book.

About the Author

LAURA LUSH TEACHES CREATIVE WRITING AND academic English in the School of Continuing Studies at the University of Toronto. Her previous books include *Carapace* (Palimpsest Press, 2011), *The First Day of Winter* (Ronsdale Press, 2002), *Going to the Zoo* (Turnstone Press, 2002), *Fault Line* (Vehicule Press, 1997), and *Hometown* (Vehicule Press, 1991). Her poems and short stories have been anthologized in both Canadian and international journals. She lives with her son, Jack, in Guelph, Ontario.